© Aladdin Books Ltd

Designed and produced by
Aladdin Books Ltd
70 Old Compton Street
London W1

Published in the USA in 1984 by
Franklin Watts
387 Park Avenue South
New York, NY 10016

Library of Congress
Catalog Card No 84-51228

ISBN 0-531-04836-5

Printed in Italy

Rivers
and Lakes
Jenny Mulherin

FRANKLIN WATTS
London · New York · Toronto · Sydney

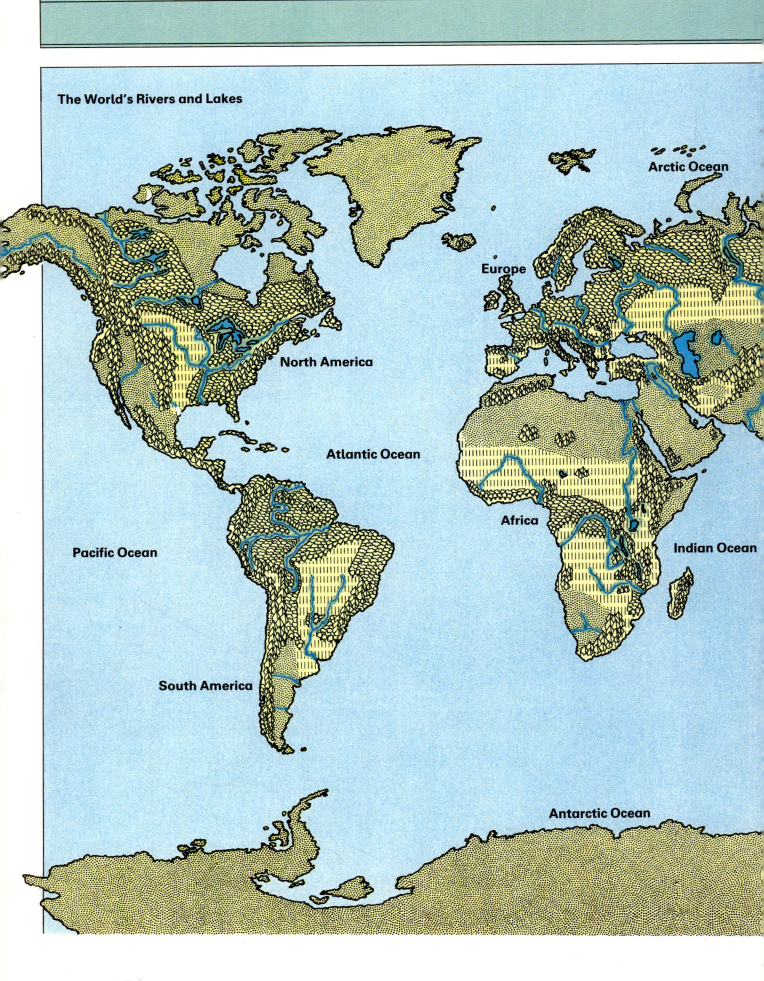

The World's Rivers and Lakes

Arctic Ocean

Europe

North America

Atlantic Ocean

Africa

Indian Ocean

Pacific Ocean

South America

Antarctic Ocean

Foreword

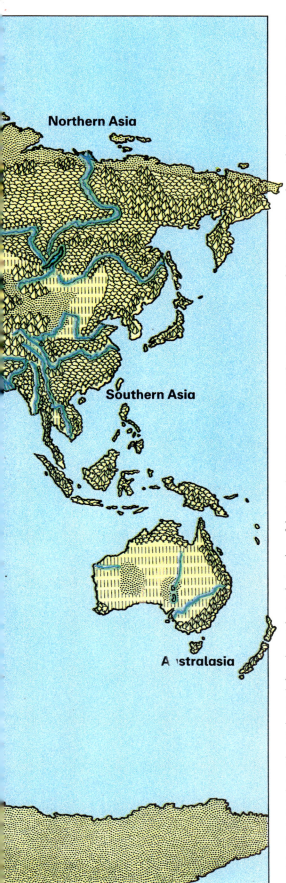

Northern Asia

Southern Asia

Australasia

Without rivers and lakes and the fresh water they provide, there would be no life on the surface of the Earth. Without rivers there would also be a very different landscape, because over millions of years they have shaped and molded great areas of the world's surface. Although rivers and lakes contain less than 0.02 percent of the world's water, this is enough to provide life and nourishment for the people, animals and plants that live on Earth.

Rivers provide, but they also have the power to destroy. They provide water to drink and grow crops, waterways along which we travel, and power to fuel industries. For example, without the Kariba Dam built on the Zambezi River and the electricity it provides, the economic life of Zimbabwe and Zambia would sharply decline. But some rivers periodically flood the land, ravaging crops and drowning people and animals. One of the worst floods in history occurred in 1933 when the Yellow River in China burst its banks, killing over 3.5 million people. Taming the great rivers and using them, rather than abusing them, has been one of the major problems of our time.

Contents

River Formation

For us, fresh water is essential. Without it most people could not survive for much more than 10 days. Our bodies actually consist of 65 percent water! And using it in every imaginable way, people, living in the West at least, account for an average 1100 liters (291 gallons) each every day – all this from less than one percent of the world's water. The rest is in the oceans, with about two percent locked up in the great ice sheets of the North and South Poles. Where then does the continual supply of fresh water come from? To answer this we must first understand that all water is part of a huge ever-moving cycle.

The water cycle
The water cycle starts when the Sun heats up the Earth's water, mostly sea water, and causes it to evaporate. It rises into the air in the form of water vapor. As the air cools high above the sea, the invisible water vapor turns into droplets or ice crystals which form clouds. As the clouds grow, they become heavier and finally all the moisture they contain is dropped on the Earth as rain, hail or snow. The water either drains into streams and rivers, or seeps underground through rocks. Both underground water and river water eventually find their way back to the sea and so the water cycle begins all over again.

How rivers are formed
Many rivers start life as glaciers, large bodies of ice which form in high mountain regions. In spring and summer, when the ice and snow melt, turbulent streams of water pour down the steep mountain valleys, to form a fast-flowing river.

Other rivers start in springs. As rainwater seeps through the soil into the

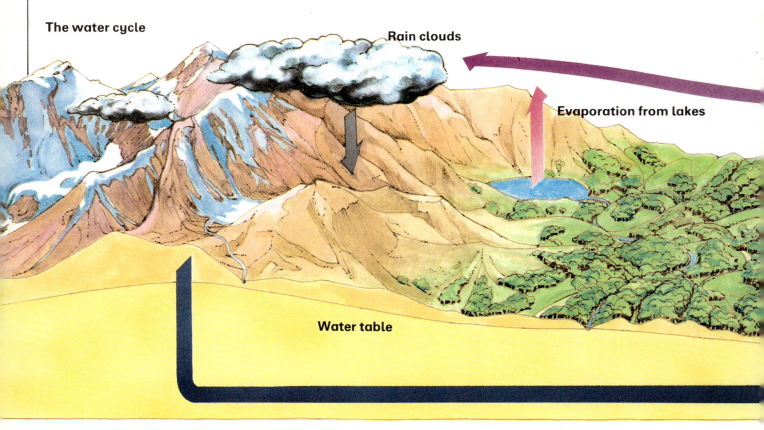

The water cycle

Rain clouds

Evaporation from lakes

Water table

rocks below, it meets layers of permeable and impermeable rock. Where this is near the surface, underground water collects and forms into a spring. Limestone is a permeable rock – it allows water to flow through it. When the water reaches the impermeable rock, it flows out of the ground. This layer is the water table.

Some rivers flow from lakes. The source of the Nile, for example, is Lake Victoria in East Africa. Heavy rain or melting snow may raise the level of a lake, which eventually overflows to form small streams. Several small streams may then join and so a river is born.

How lakes are formed

Lakes are formed in depressions or hollows in the Earth's surface. Many depressions were created by ice action that ended about 11,000 years ago. As the ice advanced, it gouged out rock basins and carried with it rocks and stones called moraine. When the ice melted, these were left as ridges. The rock basins filled with water and the moraine dammed many rivers. Other lakes collected in depressions created by the uplifting and warping of the Earth's surface, or when landslides or lava blocked the flow of rivers. In some cases, they form in the craters of volcanoes.

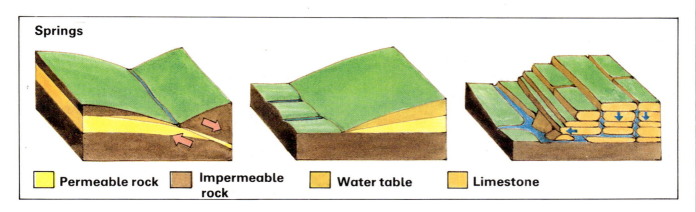

Springs

☐ **Permeable rock**   ☐ **Impermeable rock**   ☐ **Water table**   ☐ **Limestone**

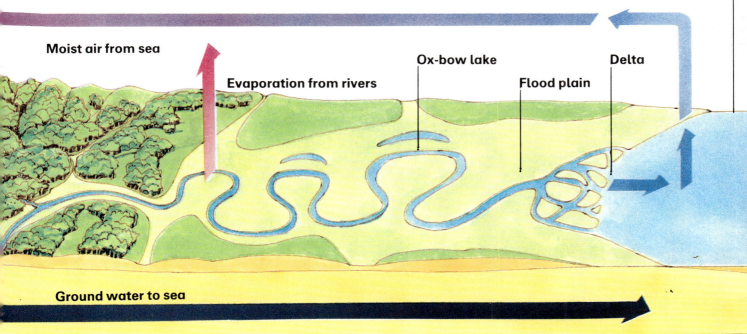

Moist air from sea

Evaporation from rivers

Ox-bow lake

Flood plain

Delta

Ground water to sea

The Life of a River

The life of a river is usually divided into four stages — infancy, youth, maturity and old age. The terms really describe the source, upland, middle and lowland parts of a river, and these parts are determined by the slopes of varying steepness through which the river flows.

In infancy, water melts off the peaks (1) of the mountains and then re-freezes into glaciers (2). As the glacier descends it melts to form lakes (3). As the lakes fill up they overflow into fast-flowing rivers (4). A river here will carry stones and rocks from the upland slopes which are mainly responsible for carving out the deep valleys (5), which the river runs through. Waterfalls (6), and rapids (7) are common in the youth stages of a river.

The mature stage begins after it leaves the mountain. Its slope is less steep than before and the river starts to swing from side to side, widening its valley. Other rivers join too, so mature rivers generally carry a greater volume of water than in their youth. They also carry a large amount of eroded material, some of which is deposited on the riverbed and banks. As it moves downstream it often develops broad bends called meanders (8) which can become oxbow lakes (9). These are crescent-shaped bends that become cut off from the main channel (10).

Old age
In its final stage the river flows sluggishly, carrying its greatest volume of water and eroded material. It flows over almost flat plains, called flood plains (11) because at this stage rivers regularly overflow their banks, depositing the fine, ground-down rock, called silt, over the land.

When the river reaches the sea, the fine silt is dropped at the river's mouth, forming a fan-shaped area of swampy land called a delta (12). The deposits at the mouth of the river (13) will eventually create more land as they build up, under water.

Infancy

Youth

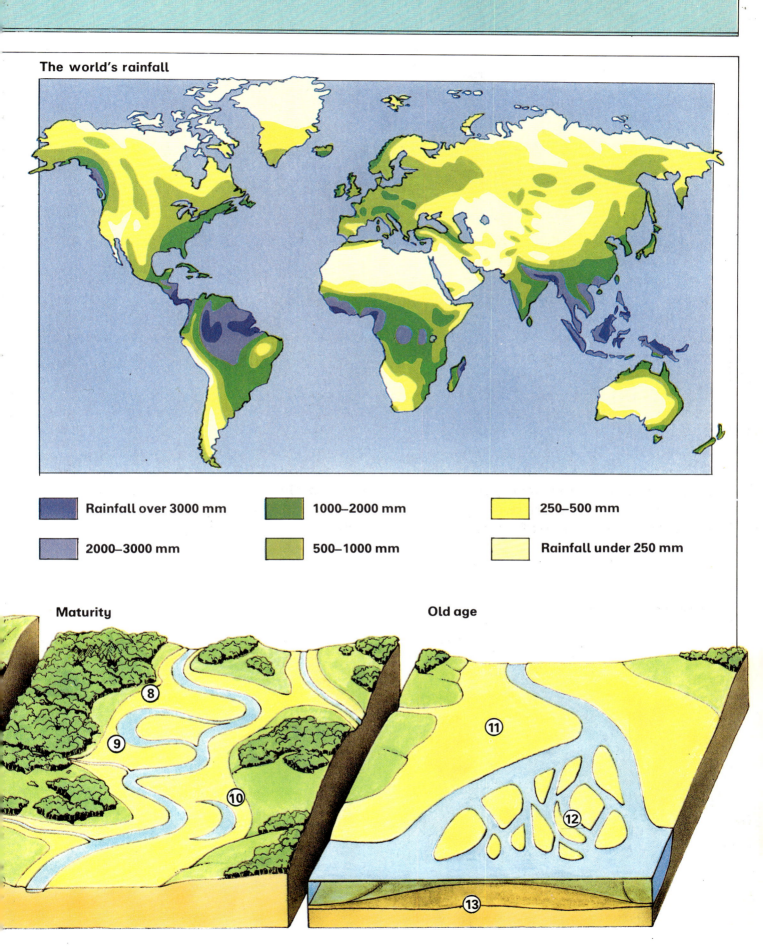

The world's rainfall

Rainfall over 3000 mm

2000–3000 mm

1000–2000 mm

500–1000 mm

250–500 mm

Rainfall under 250 mm

Maturity

8

9

10

Old age

11

12

13

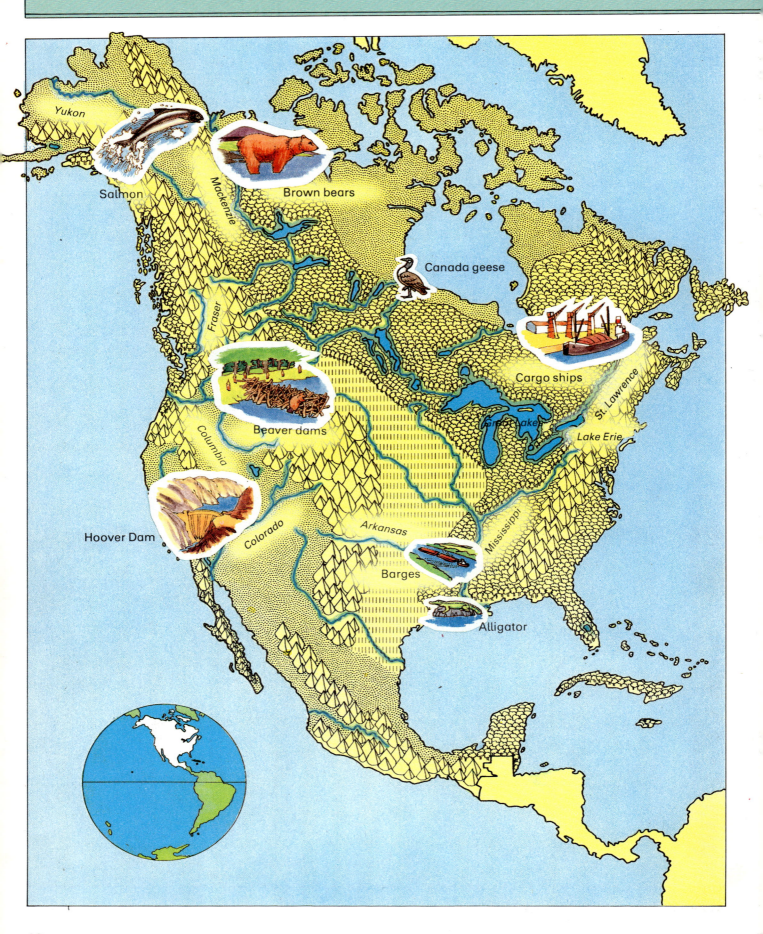

Yukon

Salmon

Mackenzie

Brown bears

Canada geese

Cargo ships

Fraser

St. Lawrence

Great Lakes

Lake Erie

Beaver dams

Columbia

Hoover Dam

Colorado

Arkansas

Mississippi

Barges

Alligator

North America

North America is a continent of immense lakes and rivers. It contains the Mississippi, one of the world's longest rivers, and the Great Lakes, which combined form one of the greatest concentrations of fresh water in the world. Most of the continent's longest rivers are east of the Rocky Mountains, including the Missouri-Mississippi, Ohio, Arkansas, and in the north, the Saskatchewan and Mackenzie. However, important rivers, such as the Colorado, Columbia, Fraser and Yukon do flow westward from the Rockies to the Pacific. The Colorado is a particularly interesting example of a river rejuvenated, or made young again. Over a million years ago, the Colorado plateau was probably a flat coastal plain across which the river flowed slowly. Earth movements then lifted the plain so that the river was forced to increase its power and cut down through the rocks. The result is the Grand Canyon.

As in Europe, a great number of North America's lakes are found in the colder latitudes. In some regions of Canada, lakes formed by ice action millions of years ago are so numerous that the landscape consists of more water than land. The Great Lakes, originally vast hollows formed by Earth movements, were later deepened and cut down by ice. As the ice melted, water gathered in them.

Colorado River, running through the Grand Canyon

In a technologically advanced country, such as the United States, rivers such as the St. Lawrence Seaway and the Mississippi are still important as routeways. The livelihood of the people of the Deep South has always been intimately connected with the Mississippi. Cotton plantations still line the river banks, tobacco has been grown here for hundreds of years, and barges still carry crop cargos downstream to the great port of New Orleans. There is an enormous volume of river traffic on the Mississippi; one of the world's most important oil refining centers is at Baton Rouge, just a little upstream from New Orleans. Great chains of barges, pushed rather than pulled by towboats, form the main traffic on the river. Barges sometimes linked three abreast carry oil, gasoline, grain, chemicals, coal and metal ore. It takes great skill to navigate the Mississippi.

Harnessing the Colorado's power

Although the Colorado passes through the least inhabited region of the United States, engineers are controlling it, with such dams as the Hoover Dam, to provide hydroelectric power, domestic water supplies, and irrigation for the people not only in the towns and desert nearby, but also for millions of people in areas hundreds of miles to the west, including the city of Los Angeles. Crops grow on what was once desert land. Cotton, citrus fruits, vines, vegetables and date palms are all crops that are now cultivated successfully, thanks to year-round irrigation.

Barge towboat on the Mississippi River

Indian fishing for salmon

Wildlife

One of the most remarkable river dwellers in North American waters is the beaver, once extensively hunted on rivers such as the St. Lawrence for its fur. With only their teeth and claws, they are able to construct dams, canals and their own dwellings known as "lodges." Beavers build dams to stop the flow of rivers in order to make artificial lakes, on which they construct their homes.

But it is in the delta of rivers such as the Mississippi that the most astounding variety of wildlife and plants is found. Here, in the backwaters or "bayous," are alligators, giant frogs and catfish. Millions of birds, such as Canadian and Lesser snow geese, mallard, teal and egrets, winter on these lower stretches of the river. The path of these birds is aptly called the "Mississippi Flyaway."

Migrating salmon

Many species of fish travel along rivers at certain times to breed. The salmon, which is found in rivers in North America and Europe, begins life as an egg on an upstream, stony riverbed. After the egg hatches into a tiny fish it gradually moves downstream where it begins to feed. After one or two years, the salmon starts its journey back to its birthplace to breed. Once at its birthplace the salmon breeds and then attempts its journey back to the sea. Few salmon make the return journey. Weak from exhaustion, they are easy pickings for predators. The fiercest of these are the salmon-eating brown bears that wade into Alaskan rivers and pounce on the swift-moving, but exhausted, salmon, catching them with their paws and mouths.

Beaver

Alligator

Snowy egrets

Inia

Capybara

Orinoco

Poison fishing

Giant
redfish

Amazon

Lake Titicaca

Aymara Indian

Parana

Rednecked
turtle

Paraguay

Piranha

Uruguay

South America

In South America the most important rivers and lakes are found on the east side of the continent. The main rivers, such as the Amazon, Orinoco and the river system formed by the Paraguay, Parana and Uruguay, flow into the Atlantic. Because the Andes are so close to the coast there are few rivers of any significance flowing into the Pacific. Most of South America's lakes are mountain lakes formed by upheavals on the Earth's surface, and are found in the Andes or along their foothills. The largest of these is Lake Titicaca, the highest lake in the world.

South America's most important river, the Amazon, cuts through narrow canyons from its source in the Andes. Along its 6400 km (4000 mile) course, it collects more than 1000 tributaries. In its lower course, it snakes through the flatlands of Brazil which it floods regularly, sometimes for as much as 96 km (60 miles) on either side. The Amazon's gradient is so low that tidal waters sometimes extend about 965 km (600 miles) inland from the Atlantic. Further south, the Paraguay, also a river of the plains, flows through a wide stretch of marshes (the Pantanal) in its middle course. By contrast, the Parana flows mainly through the high plateaus, while the Orinoco plunges down steep slopes before it too flows along a flat basin.

The Amazon River

People

Life on the lakes

Compared with North America, relatively large numbers of South Americans depend directly on rivers and lakes for their livelihood. The Aymara Indians live on Lake Titicaca, 3,800 m (12,500 ft) above sea level in the Andes.

Like their ancestors, many of these people still live on floating islands of dried tortora (a reed like papyrus that grows in marshy shallows) which are anchored with rocks and frequently bolstered with more reeds. The floating reed island often houses two or three families. Their main sources of food are carp and catfish, which they dry in the Sun and then smoke. The men fish from boats made of bundles of dried reeds lashed together. These look remarkably like the papyrus boats pictured in the painted tombs of ancient Egypt.

Nowadays, the lake dwellers no longer live by fish alone; the cutting of reeds where the fish breed has depleted the lake's resources.

Life on the rivers

Tribes living along the Amazon fish by hook and line, bow and arrow and by poisoning. Equipped with timbo poison from the bark of a tree, fishermen dam streams with bamboo and reeds, and poison the lake below the dam. After a short while, the paralyzed fish float to the surface to be scooped into baskets by the waiting Indians.

Poisoning the river for fish

Indian fishing with bow and arrow

Wildlife

A rather spectacular inhabitant of the Amazon is the giant redfish or pirarucu, one of the largest of all freshwater fishes. Some grow to a length of 5 m (16 ft) and weigh about 23-27 kg (50-60 lb). This massive fish is greenish at the head and bright red toward the tail. Tiny piranhas, however, are the most dangerous fish in South American rivers. Although their usual food is other fish, piranhas will often attack any animal that enters the water. Their teeth are so sharp and their jaws so strong that they could snap off a finger in one bite. It is said that a group of piranhas can reduce a capybara weighing 45 kg (100 lb) to a skeleton within three minutes.

The capybara is the largest of all rodents and is a sort of gigantic, water-loving guinea pig, commonly found in the rivers and lakes of Central and South America, as is the rednecked turtle.

Aquatic mammals are generally found only in salt water but a dolphin-like creature called the inia lives in the waters of the Amazon, Orinoco and their tributaries. Usually 2 m (6 ft) long, inias feed mainly on fish, and like all dolphins and whales they come to the surface to breathe. The Indians regard the inia as sacred and never hunt it .

There is much exotic flora on the South American rivers. One flower which grows in profusion is the water hyacinth. Its small leaves and pretty blue flowers grow with amazing speed, choking rivers — it has become a pest.

Red piranha

Water hyacinth

Capybara

Hydroelectric power

Wildfowl

Lake Ladoga

Kingfisher

Rhine barge

Water vole

Eels

Brown trout

Vistula

Loire

Marsh marigold

Giant catfish

Danube

Crayfish

Europe

Most of Europe's rivers have their source in the continent's many mountains and hills formed millions of years ago, although some mountain ranges have actually been cut through by rivers. The Danube, for instance, cuts through the Carpathian mountains at several points – notably the "Iron Gates" in Romania. Mature rivers are found in western, central and eastern Europe; their valleys slope downward gradually and they are navigable. Many of them, such as the Rhine and the Volga, are important waterways. In European Russia these rivers are long and flow sluggishly across flat fertile plains. By contrast, younger – shorter and faster – rivers are found in northern and southern Europe, and such rivers are very suitable for producing hydroelectricity. Norway and Sweden, for example, obtain more than three-quarters of their electricity from hydroelectric stations.

Lakes cover only about two percent of Europe's surface. Most of them are found in northern Europe and Scandinavia. In Finland, for example, there are about 60,000 lakes which make up about one-fifth of the entire country. These northern lakes were made by ice action but other European lakes have been formed in depressions dating from long before the Ice Ages. Lake Geneva, and Lake Ladoga, which is the largest lake in Europe, were formed in this way.

Norwegian fjord

People

The Rhine is the busiest and most commercially important river in western Europe. It is also one of the dirtiest, largely because it flows through the industrial heartland of Europe, the Ruhr, a vast region of coal and steel. Huge chimneys pumping out great quantities of smoke, enormous factories, warehouses and blast furnaces line the river's banks for almost 30 km (48 miles) from Dusseldorf to just beyond the large industrial city of Duisberg. From here thousands of huge barges carrying up to 2000 tonnes (1970 tons) travel down the river to the sea.

Almost all the fish that lived in this part of the river have died. Salt pollution from France and phosphates associated with detergents are two of the major problems; in 1969 large quantities of a deadly pesticide, endosulfan, fell off a barge into the Rhine, killing millions of fish. Now attempts are being made to clean up the Rhine.

A river port

Wildlife

Almost all the rivers in temperate regions of Europe contain a common variety of wildlife, but the plants and animals found near a river's source are different from those in its middle course and toward its mouth. Geographers have divided European rivers into five stages according to the wildlife contained.

Headstream and troutbeck
At its source, the highest part of the river, the water is clear, cold and fast-flowing. Relatively few plants and animals can live in this environment and the river is mostly inhabited by insects such as stoneflies and mayflies. Just a little further downstream the troutbeck still contains fast-moving water. There are very few plants here

Water vole

but as its name suggests, brown trout are found here in rock pools.

Minnow reach

In the minnow reach the river is wider and flows more slowly. Wildlife becomes richer in this section of the river. Marsh marigold and loosestrife are among the many plants, while crayfish and insects provide food for fishes such as eels, grayling, minnow, salmon and trout. Young eels and other fishes of the minnow reach attract birds like the colorful kingfisher.

Lowland and estuarine reaches

In its lower reaches, the water, now warmer and rather murky, flows more slowly. This section of the river is rich in wildlife. Water crowfoot and Canadian pondweed flourish along the river bank. Bream and perch are found here as well as pike, which live on other fishes. Young water birds and water animals such as the water voles live on fish and plants. Finally, in the estuarine reach, where fresh water mixes with sea water, the bottom-feeding flounder is common, and herons, gulls and wildfowl feed on the mud flats.

One European river which is particularly rich in natural life is the Danube, where over 70 different kinds of fish are found. This is because the river is a link between the fish species of western Europe and Russian freshwater fish. It has two special kinds of pike-perch, its own species of salmon and a giant catfish which grows up to 5 m (16 ft) in length.

Kingfisher

Marsh marigold

Gray heron

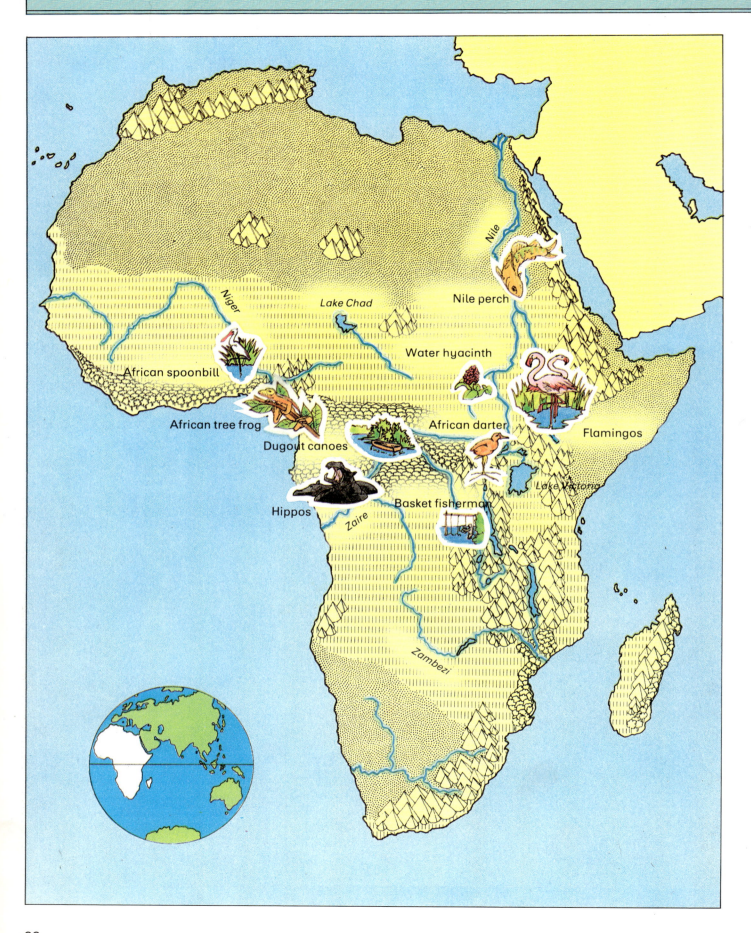

Niger

Lake Chad

Nile

Nile perch

Water hyacinth

African spoonbill

African tree frog

Dugout canoes

African darter

Flamingos

Hippos

Zaire

Basket fisherman

Lake Victoria

Zambezi

Africa

All the major rivers and lakes in Africa have one factor in common. Their origins lie in the gigantic upheavals in the Earth's crust that created ridges, plateaus and depressions on the continent's landform. All important African rivers have rapids, cataracts and waterfalls, caused by ridges forming across the course of a river, or by bands of hard rock which the river's flow has not worn away.

Lake Chad in central Africa and the lakes of East Africa, such as Lake Victoria and Lake Eyasi, are economically very important to their surrounding countries. They are also the home for many remarkable species of plant and animal life, some of which are unique to Africa.

Africa's four major rivers are the Nile, the world's longest river, the Zaire (formerly the Congo), the Zambezi and the Niger. The Nile flows from Lake Victoria north to the Mediterranean, taking in many tributaries and flowing over many waterfalls and rapids. Its course is also interrupted by swamps, the largest of which is the Sudd, a vast area of floating swamp reeds, mostly papyrus. From here it runs through 1000 km (620 miles) of desert until it reaches the delta. The Zaire also has many waterfalls and rapids but its course often runs through deep gorges before finally cutting a narrow exit into the Atlantic.

Victoria Falls

Wildlife

Africa's lakes and rivers support an extraordinary variety – and a remarkable profusion – of wildlife. The more exotic animals occupy the tropical belt running across the center of Africa. This is the home of the hippopotamus, the largest water-dwelling animal in the world, which is especially common along the banks of the Zaire. Because they cannot sweat, hippos have to control the temperature of their bodies by staying in the water. When darkness falls and it is cool, the hippos leave the water, and go into the forest in search of food. Although they are such huge animals, hippos live entirely on grasses and leaves.

Despite hunters, crocodiles still flourish along the Niger and the Nile rivers.

The Nile crocodile which grows to a length of 6-8 m (20-28 ft), looks ferocious, but is usually content with a diet of small mammals and fish. One of the largest fishes in African rivers is the capitaine or Nile perch. This gigantic fish can weigh up to 90 kg (200 lb), is highly prized and brings a good price when fishermen sell it to river traders.

Birds abound on African rivers and among the largest are the African spoonbill, which sifts food from river mud, and the Sacred ibis. Another bird, the African darter, actually swims in rivers with only its head and neck showing. Among other interesting river dwellers is the African tree frog which builds its nest on a branch overhanging the river. It beats

Greater flamingos

Marbled reed frog

Hippos

its egg jelly into a foam which hardens on the outside. Eventually, the tadpoles drop into the river.

Life in Africa's lakes

The animals that live on or around lakes are generally much the same as those that live on the rivers that flow into them. In Africa, however, some birds have adapted to the special conditions of African lakes. Flamingos, for example, are the only birds to raise their young on the soda lakes of East Africa. Flamingos feed by using the fine hairlike filters on the edge of their bills to sift out algae from the poisonous water and mud. In African lakes, as elsewhere, the water hyacinth has become a pest.

The Aswan Dam was built on the Upper Nile in the 1960s. Behind it stretches a reservoir 480 km (300 miles) long. Its main purpose is to supply hydroelectricity for Egypt's industry, but it has also helped to irrigate huge areas of once barren desert. However, lower down the Nile traditional irrigation methods persist. The shaduf is simply a bucket or a weighted pole used to raise river water and pour it into ditches behind the bank. The Archimedes Screw raises the water by means of a rotating spiral set in a cylinder. The base of the cylinder rests in the river, and as a handle is turned, water is drawn up the spiral to pour out at the top.

Fishing methods vary from place to place on the African rivers. Around the cataracts of Stanley Falls on the Zaire, the natives employ a complicated system of baskets set on vertical poles stretching well into the river. The baskets are winched out of the water at regular intervals, and the catch removed. This is a very effective method of fishing.

Fishing around Stanley Falls

Northern Asia

Northern Asia is a land of many rivers and lakes, most of them formed in northern regions by ice action. The great Siberian rivers, such as the Ob, the Yenisei and the Lena, all flow into the Arctic Ocean and freeze over in winter. Although they flow through inhospitable country, these rivers are of vital importance for the people who live in these Arctic regions. They are routes of communication used by boats in summer and in winter they become roads for sleighs.

Thousands of small lakes formed by glacial action are also found in these Siberian regions. Other lakes in northern Asia, such as Lake Baikal, the deepest lake in the world, are economically important because of the large numbers of fish they contain. This lake is remarkable — because it is so clear and rich in oxygen, plants are able to grow as far down as 67 m (220 ft) and animal life flourishes near the bottom of the lake. About 300 of northern Asia's streams and rivers flow into Lake Baikal.

The two great rivers of eastern Asia, the Huang Ho (Yellow River) and the Yangtze Kiang flow in long, winding courses into the East China Sea. The basin formed by these rivers is known as the China Plain, a vast area of fertile land which is intensively farmed and densely populated. These two rivers are noted for the disastrous floods they have caused in the past. Only now are floods being brought completely under control.

Huang Ho

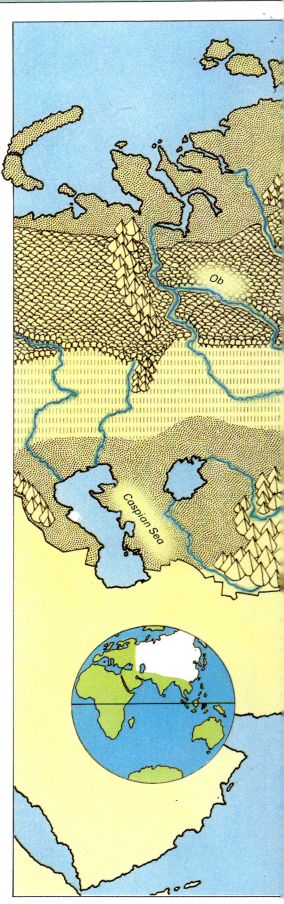

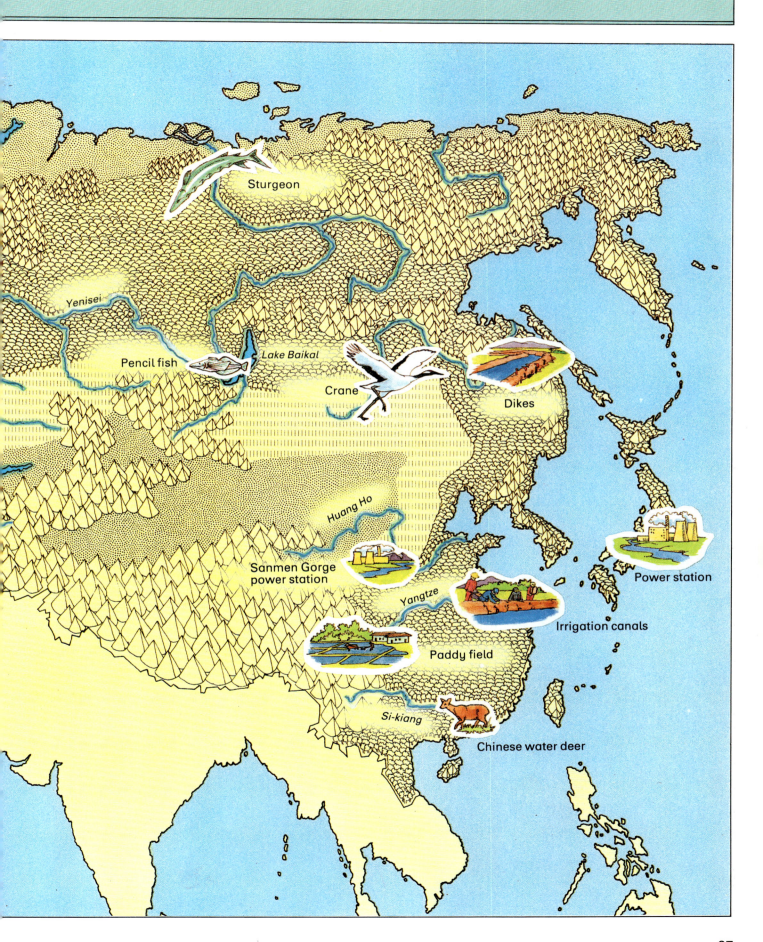

Sturgeon

Yenisei

Pencil fish

Lake Baikal

Crane

Dikes

Huang Ho

Sanmen Gorge
power station

Power station

Yangtze

Irrigation canals

Paddy field

Si-kiang

Chinese water deer

The Yellow River is one of China's most important rivers and also one of its most destructive. Named the Huang Ho in China, it is also called the Yellow River because of the large amounts of sediment it carries. In spring and summer, melting ice near the river's source coincide with heavy monsoon rains. This swells the river, and great waves of water move rapidly downstream. Unless strong walls or dikes have been built, the river is likely to break its levees and flood the plain.

Taming the river

But an ambitious scheme is well under way to control the Yellow River. The Yellow River Commission, set up to examine ways in which flooding could be prevented and to see if its waters could be used for irrigation and hydroelectricity, suggested the construction of 46 large dams on the river's middle section and proposed that much of the silt on the river bed be removed. The Sanmen Gorge project is the largest of the 46 proposed dams. It holds back the river for nearly 300 km (186 miles) and its reservoir covers an area of more than 2000 sq km (722 sq miles). The dam not only helps prevent floods in the river's lower reaches but it also provides water for irrigation and hydroelectricity. Schemes such as this are also being carried out elsewhere as in Japan.

Irrigation has led to the cultivation of rice in paddy fields in large areas around the rivers. Rice is the staple food of the people of southern North Asia. Rice needs to be submerged under water during its growing season, but before this can take place the land must be drained and the soil prepared for the crop. So it is important to be able to both drain and flood the fields at different times.

Japanese dam

Paddy fields

Wildlife

The wildlife of North Asia's rivers and lakes varies according to region and climate. In Lake Baikal plant and animal life is especially interesting because much of it is found nowhere else in the world. One unique species of fish is the Comephoridae. These are the size of a pencil, have transparent bodies without scales, and extremely pure fat that makes up about 25 percent of their weight. When they are taken out of the water and left in the Sun, these fish literally melt away. The Baikal seal, or the ringed seal, is the only freshwater seal in the world. Almost destroyed by fur hunters in the past, it is now a protected species.

One of the most important fishes found in many of the rivers and lakes of northern Asia is the sturgeon, highly valued for its roe, known to us as caviar. In Siberian waters, this fish can measure up to 3 m (10 ft) long, and weigh about 230 kg (500 lb).

Birds and animals

One unusual creature, the Chinese water deer inhabits the swampy rivers of south-eastern China. A small animal without horns, it has long canine teeth like tusks. Its home is in marshes and it swims perfectly. It looks rather like a pig and like pigs it bears as many as seven young, instead of one or two like all other deers. Historically revered by the Japanese is the Bird of Happiness. This is the red-crowned crane which stands 1.8 m (5 ft) tall and lives in China, Korea and Japan. The crane has long been worshipped as a symbol of happiness, longevity and fertility. These cranes can live for more than half a century and they have only one mate for the whole of their lives.

Sturgeon

Chinese water deer

Ringed seal in Lake Baikal

Southern Asia and Australasia

Many of the rivers of southern Asia are young rivers which have short and usually violent courses. The Mekong, although a long river, descends quickly through spectacular mountains and gorges before passing by the plains of northeast Thailand and Kampuchea to its delta in southern Vietnam. Unlike many great rivers, the Mekong is not a major highway; for much of its course, it flows through remote, barely explored countryside as yet unharnessed by dams despite ambitious plans to control the river. By contrast, the Ganges of India and its mighty tributary, the Brahmaputra are, for much of their courses, slow, sluggish rivers carrying vast quantities of silt to form a huge delta.

One of the most striking features of many southern Asian rivers is the load of mud they carry. This is caused by the heavy and often unpredictable rainfall that falls in these regions, the turbulence of many of the rivers, and the shallow seas into which they flow. As a result, large deltas, often fertile and densely populated, are a feature of many rivers, including such giants as the Ganges-Brahmaputra, the Mekong and the Irrawaddy.

Many young rivers in Australia and New Zealand provide valuable hydroelectricity, and in Australia the two major rivers, the Murray and the Darling, water the otherwise arid lands to the west of the continent's Great Dividing Range.

Mouth of the Indus

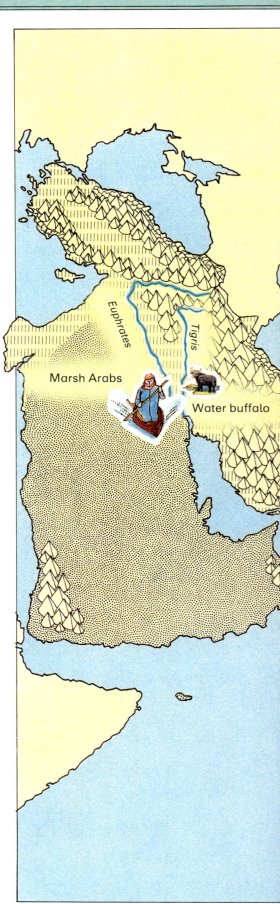

Euphrates

Tigris

Marsh Arabs

Water buffalo

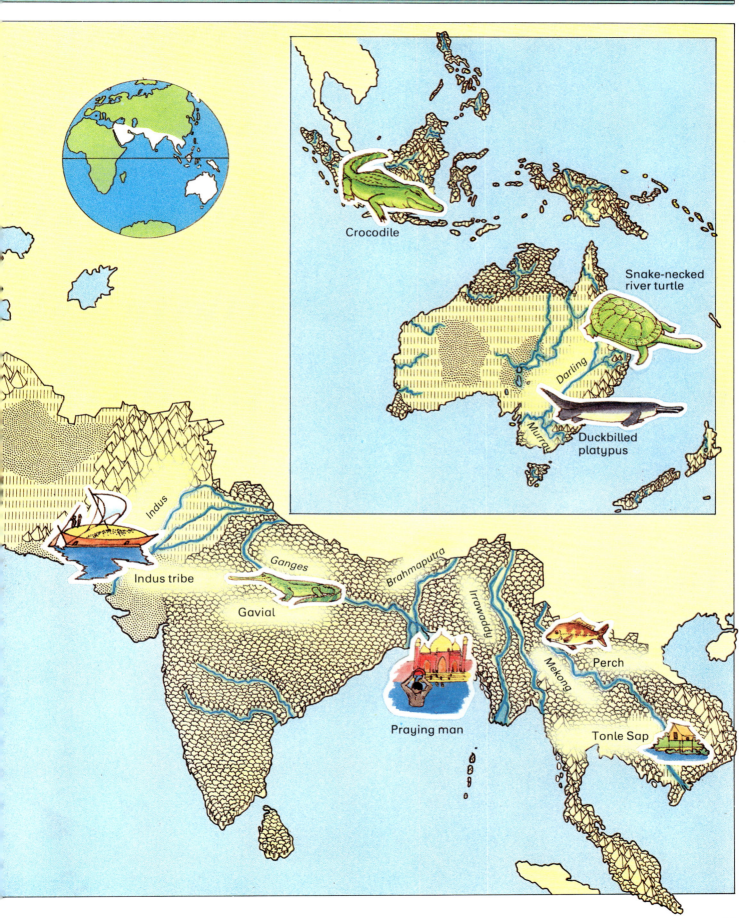

Crocodile

Snake-necked
river turtle

Duckbilled
platypus

Darling

Murra

Indus

Indus tribe

Ganges

Gavial

Brahmaputra

Irrawaddy

Perch

Mekong

Praying man

Tonle Sap

Wildlife

Some remarkable animals live in the swamps and banks of southern Asian rivers. One of the most extraordinary is a fish that is at home on land as well as in water. The climbing perch has developed breathing organs above its gills, which act as air reservoirs rather like lungs. This means that it can live on land. Another strange creature of these areas is the snake-necked turtle. This turtle has a very long neck that enables it to breathe above water while its body is submerged.

Cormorants are frequently seen in southern Asian waters. Their remarkable fishing dives are famous and this skill has been made use of by fishermen. They tame the birds and put an ivory or bamboo ring around the bird's neck so that it cannot swallow the fish it catches.

Living fossils

Another interesting creature is the gavial, a crocodile found in the muddy waters of the Ganges and Bramahputra that can grow to a length of more than 6 m (19 ft). The gavial, unlike the crocodile found throughout southern Asia, does not attack man and lives almost entirely on fish. Scientists believe that this curious reptile is a replica of a crocodile found in Australia over 200 million years ago, of which only fossils remain.

The duckbilled platypus is another living fossil, found mainly in the western areas of Australia. It is quite unique in the animal world as it lays eggs like a reptile but feeds its young as mammals do. The platypus makes its nest in the banks of rivers and feeds off the river life.

Climbing perch

Cormorants

Gavial

People

Villages on stilts are a common sight on Kampuchea's Great Lake, the Tonle Sap. The Great Lake, normally covering about 520 sq km (200 sq miles) but only 1.5 m (5 ft) deep, spreads to 15 times its area in the rainy season. As the water rises, villagers dismantle their homes and move. When the water subsides they move back, for these are among the richest fishing waters in the world. Their diet of fish is supplemented by storks.

The people that live on the River Ganges regard the river as sacred. They drink the water and bathe in it as well as pray to the fast-moving waters. When they die, their bodies are placed on wooden rafts or pyres which are then set alight. They are then pushed into the current of the river to float away downstream to honor the gods.

Marsh dwellers
Some of the most picturesque river people are the Marsh Arabs of Iraq, who live in a vast area of lakes and marshes where the two great rivers of ancient times, the Tigris and Euphrates, join before they enter the sea. These people build their houses on artificial islands made from layers of reeds. Their lofty, elegant houses are also made entirely of reeds. Water buffalo share the reed islands with families and are rarely killed or eaten. They provide milk and also fuel for fires since their dung is dried into round flat cakes and then burned. The people of the Indus River are very similar to the Marsh Arabs, depending upon the water for their livelihood. Their staple diet consists mainly of rice and fish.

Tonle Sap

Praying in the Ganges

The Future of the Environment

It is now realized that when people alter the balance of natural water systems problems arise. Egypt's Aswan Dam, completed in 1971, doubled Egypt's electrical power and improved more than one million acres of land. But, because it is trapped behind the dam, fertile Nile silt no longer enriches the land and farmers have to use expensive fertilizers. Another consequence of the dam, bilharzia, has alarmed health experts.

Mekong Development Project

Some 28 million people live in the lower basin of the Mekong River, and their hopes for prosperity lie in the Mekong River Development Project, a vast international enterprise costing millions of dollars. These people hope to control the river and provide hydroelectric power by building dams along the Mekong. However, the project was seriously hampered by the Vietnam War and the continued guerrilla fighting in Kampuchea. Although several dams have been built on the tributaries of the Mekong, in Thailand and Laos the giant Pa Mong Dam has yet to be built.

It would be one of the biggest dams in the world. Nearly 100 m (325 ft) high, its reservoir could irrigate about 2 million hectares (5 million acres) in Thailand and Laos. But like the Aswan, it could flood the homes and lands of thousands of farmers who would have to move to other areas. It could also change the habitat of fish in the Mekong; the giant catfish, which grows to a length of 2 m (6 ft), could die out.

Dam on the Arkansas River

Polluted river in France

Rivers and international disputes

Difficulties always seem to occur when a river flows through more than one country. Although the Ganges and the Brahmaputra rise in India, they meet in Bangladesh where each year their floods destroy crops, homes and people. Controlling the rivers to produce power and regular water supplies for irrigation, beneficial to all, requires cooperation, which so far has been lacking.

Mutual mistrust and religion differences mean that Bangladesh may continue to suffer floods and destruction. In North America, the massive North American Water and Power Alliance (NAWAPA) scheme, involving the US, Mexico and Canada, is troubled by similar problems. This scheme proposes that some of Canada's rivers be diverted south to supply water not only to the semi-arid areas in the western US but also to Mexico. The scheme also proposes to link the west coast to the Great Lakes by a series of navigable waterways.

Pollution

Pollution is one of the major problems of rivers and lakes in industrialized countries. On North America's Lake Erie, for example, 87 beaches were closed down in 1968 because of water pollution, and the sale of fish from the lake was restricted. Many other lakes and rivers have been similarly affected but, thanks to conservation efforts, large amounts of money are now being spent to clean up the rivers and lakes.

Interesting Facts

The true source of the Amazon was only discovered in 1953. It is a stream named Huarco which rises near the summit of Cerro Huagra, 5238 m (17,188 ft) high in Peru.

Between 1960 and 1963 over 10 million fish were killed in the Mississippi by a poisonous chemical that leaked into the river.

Victims of drowning in Lake Baikal vanish without a trace. Their flesh is quickly eaten by crustaceans and their bones dissolve in the almost mineral-free water.

The Tonle Sap River in Kampuchea is probably the only river in the world to reverse its course twice a year. During the rainy season from June to October, it flows from the Mekong River into the Great Lake, and from November to May it flows in the opposite direction toward the Mekong.

The flood plain of the Sudd, the great swamps of the Upper Nile, is as large as Belgium, the Netherlands and Switzerland combined.

The elephant trunk fish, an electric fish from West Africa, is now monitoring pollution in the waters of some German cities. The fish senses pollutants in the water and emit an electrical signal which sets off a warning light or buzzer, informing scientists of chemicals in the water.

The Huang Ho (Yellow River) in China has changed its course at least 26 times in recorded history.

The water at the Niagara Falls drops at the speed of 80 km (50 miles) per hour.

The 362-km (225-mile) Jonglei Canal Project, begun as a joint effort by Egypt and Sudan in 1976, will divert the waters of the White Nile by bypassing the swamps of Sudd. It will provide a constant supply of water for the people and their livestock throughout the dry season.

The Caspian Sea, the world's largest lake, is about five times as salty as the oceans.

The world's first tidal dam was built at the mouth of the River Rance in France. The flow through the dam drives turbines that generate electricity.

Species of freshwater seals live in Lake Baikal and in one or two Finnish lakes. It is thought that they were cut off from the sea about 8000 years ago.

In Rwanda in Africa, papyrus, found in many swamps and rivers, is being compressed into briquettes. It is an excellent substitute for charcoal, produces little smoke and has a low ash content.

The 2000-year-old Chinese Dujiangyan irrigation system on the Min River, a tributary of the Yangtze Kiang, is still in use today.

The world's largest delta is that of the Ganges and Brahmaputra in Bangladesh and West Bengal. It covers an area of 75,000 sq km (30,000 sq miles).

Lake Baikal

Glossary

Scientists use many special words to describe rivers and the wildlife found in rivers. These are some of the most common ones you are likely to come across.

Algae Family of simple plants', including seaweed, almost all of which live in water.

Archimedes Screw A device rather like a gigantic corkscrew, used for lifting water.

Canal An artificial waterway often built to link two rivers together, to channel a river's water for irrigation purposes, or to bypass waterfalls or rapids.

Canyon A large narrow valley bounded by steep slopes.

Dam Usually a man-made barrier across a river built in order to control the flow of water by creating a lake or reservoir behind it. Forces of nature such as landslides or glaciers can also produce dams, as can animals such as beavers.

Delta A fan-shaped area of flat land at the mouth of a river, made up of deposits of silt.

Earth movements The uplifting and warping of the Earth's surface produced by forces inside or below the crust of the Earth.

Erosion The natural wearing away of the Earth's surface, principally by water.

Flood plain A flat area of land bordering a river which has been formed by deposits of silt carried down by the river.

Geology Study of the formation and structure of the Earth's crust.

Glacier A mass of ice that moves down a valley.

Gorge A deep narrow valley similar to but smaller than a canyon.

Gradient The degree to which the land slopes.

Headwater The part of a stream or river near its source.

Hydroelectricity Electricity produced by water power.

Impermeable rock Rock such as granite that does not allow water to pass through it.

Irrigation The artificial distribution of water on the land in order to grow crops, where otherwise there would not be enough water for agriculture.

Levee A raised bank consisting of sediment dumped by rivers when they flood. Man-made levees are often constructed to help keep the water of such rivers as the Mississippi in its proper channel.

Meander A curve in the course of a river which swings from side to side in wide loops as the river flows across the flat country.

Monsoon rains Rains in Asia carried by seasonal winds and which fall at the same time each year.

Oxbow lake A lake formed when a meandering river cuts through the narrow neck of a bend.

Permeable rock Rock which allows water to soak into it.

Rejuvenation Being made young again.

Sediment Tiny fragments of rock carried by the river.

Silt Fine material consisting of soil and very small particles of rock deposited by a river.

Tributary A stream or river that flows into the main river.

Water cycle The circulation from the oceans to the air as clouds, then as rain when it falls on the land where, as rivers and ground water, it flows back to the sea.

Index